12 Steps to Intelligent Process Improvement (IPI)

Quick Reference &
Companion to Feng Shui for Intelligent Process Improvement

Written By V. Quan Lee & J. Kay

Published by IPI LLP a subsidiary of Ying Yang Enterprises
Atlanta, Georgia

Bringing peace and harmony to the work place through
Intelligent Process Improvement

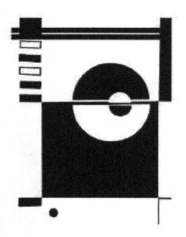

Intelligent Process Improvements (IPI) LLP
P.O. Box 813474
Atlanta, Georgia 30081
www.intelligentpi.com
vqlee@intelligentpi.com
Jkay@intelligentpi.com
Tel: 888-6yin-yang (1-888-694-6926)

What People Are Saying About Our Books:

I really like the idea of using something like (Feng Shui) that people today have had such interest in as the 'avenue' to get them to pay attention to process improvement.
Liz Woroniecki, Director of Curriculm and Training

It looks good, interesting concept and I think everyone in IT can relate.
Shariah Abdul-Karim, Senior Business Analyst Consultant

A book that bridges the gap between processes and proper methods, which uses Feng Shui to provide that balance in an employer's life and work to create the success of a project of any company.
Hillary Thomas, Software Developer/Business Analyst, Citigroup

I am not used to reading a technical book and I feel like someone is talking to me. It is refreshing! I didn't know you knew so much about Feng Shui. As an Asian, I do know a little bit about the subject and what you have in the book is accurately to the point. I like the way you're comparing Feng Shui to IPI.
Hubert Chan, Configuration Manager

It felt good to read that a graduate of Information Science and Systems is doing well. I have seen the preview of your book and it looks very good! I plan to have the MSU Library order the book both as a source of knowledge as well as a showcase of work by a fellow MSU student. I have also visited your website and it is very impressive! We feel pride in your accomplishments and wish you the very best in all your future endeavors
Jigish Zaveri, PhD
Acting Chair & Associate Professor
Morgan State University
Department of Information Science and Systems
Earl G. Graves School of Business and Management

This is the Intelligent Process Improvement (IPI) Quick Reference Guide. This is a pocket size condensed version containing the concepts that will radically change your processes and will bring harmony and balance to your work environment and projects. Not only is it fast and easy reading, it is practical and can be applied to any environment immediately.

Step 1

We agree that Process Improvement is for the people and by the people – that processes should bring harmony and balance to the workplace.

Step 2

Look at all of the best practices, methods and models that have been tried and then define a hybrid one that is unique and works best for your environment. Do not confine yourself to only what exists today.

Step 3

Make Process Improvement a legitimate project with a budget, timeline, tools (i.e. Process Portal), assigned resources and a Project Manager.

Step 4

Take full inventory of your 'as-is' processes and don't be afraid to punish anyone for admitting their weaknesses, wrongdoings, shortcuts or mistakes.

Step 5

Draw a picture for everyone to see, make a before 'as-is' picture of your processes that are visible to everyone and then use it to highlight and chart your progress.

Step 6

Continuously indoctrinate and infuse
process improvement in the culture
through training and visual displays.

Step 7

Give the Process Champions Committee (PCC) (i.e. your process improvement steering committee) centralized power to be the governing body, the police, the arbitrator, and auditors of every process and every project.

Step 8

Enforce that everyone including Executive Management is accountable to the Process Champions Committee (PCC).

Step 9

Make your improvements in short iterations. Don't try to do it all at once. Processes should be defined and deployed in 30 day intervals for major improvements and 2 week intervals for minor improvements. (FFMM)

Step 10

Keep your processes and methods flexible and agile so that you won't be so rigid that you lock yourself out of business opportunities. (Agile SDLC)

Step 11

Set quantified measurements for every
goal that you wish to achieve. Set goals
that include measuring harmony and
balance not just dollars and deadlines.

Step 12

Purge processes and avoid process overload. Never hold onto processes out of tradition.

IPI Mission Statement

The purpose of Intelligent Process Improvements (IPI) is to intelligently select and implement incremental improvements that measurably enhance the organization's processes by creating harmony and empowerment for the *REAL* workers in IT.

At the end of the day....IPI is about taking the frustration out of the IT Software Development Process.

Seven Critical Principals of IPI

Intelligently

IPI believes that intelligently made decisions are made by the people who do the real work. The people who do the real work have the insight and visibility to the lower level trenches where the heroic efforts - blood, sweat, tears (long hours and overtime) are shed to get the work done. Upper management is 'Upper' as in from the top looking down. They are looking at the big picture from a strategic point of view. Upper Management and academics only know the job in theory yet they are the ones that have the leisure to create books and methodologies that get accepted as best practices in the industry. The people doing the 'real work' have the hands on, practical, first hand knowledge but they are the least likely to write the books.

Companies are run as if only Upper Management can make intelligent decisions. IPI thinks everyone can make intelligent decisions. Intelligent decisions can only be made about that which you have intimate knowledge of. Upper Management can make decisions about the big picture because they are being feed with forecast and trends. The people doing the real work are the best resources for making intelligent decisions about what they know and what they do. IPI empowers the people doing the real work with the process improvement decision making capabilities.

IPI thinks intelligence is acquired through on the job experience not simply bought from universities. The more opportunities people get to make decisions, the better they get at it. IPI environments are foremost to promote making sound decisions based on hands-on experience.

The people who do the real work know the best approach but are often stifled. IPI refers to the old approaches to the Process Improvement top down as Upper Management dictating the processes and pushing them down upon the workers. The workers don't have input, they feel the processes don't belong to them and that the process belongs to the executives. They don't buy into the processes, they perform them because they are made to do so, and they often don't know what the process is supposed to be accomplishing. In turn, management utilizes processes as a weapon to control the so-called 'lazy' workers. The entire process implementation puts a divisor between the workers and management.

IPI challenges the 'Executives knows best' approach to process improvements and instead has the people identifying the areas of process improvement, submit the suggestion, prioritize and define the metrics, and return-on-investment. The bottom up approach to implementation process is when the users define the process and the executives are perhaps the last to know. IPI is the people educating Upper Management through processes.

From our very first implementation, we recognized that there are two groups that significantly impact the success of process improvement, the executive leadership management team and the people doing the real work. We've seen where process improvement can actually drive a wedge between these two groups. IPI couldn't miraculously pull these two groups together overnight, but instead we wanted to bring about a balance of power and communication thus creating harmony.

The best way to make intelligent decisions is knowing your options. Therefore, IPI presents a plethora of options. The most intelligent thing an executive can do is to surround themselves with the right people and then trust them to advise the executive. The one thing that the CIO on our first assignment did right was to tell us NOT to present two options but for us to go and figure out what was the one right choice and to present it to her. She made us do the research and make the uniformed decision. That was the best decision she could have ever made.

By the time an executive knows that something in the trenches isn't working, they are re-acting to the problem rather than being pro-active to the problem. The only way that the executive is even aware that something isn't working is when they are getting a status report or have gotten an email or phone call to escalate something that may have been festering or just the manifestation of something that's been an issue for some time that the workers have been aware of but had no way of implementing a change. IPI is about pro-active process improvement.

One executive we worked with would take the customer satisfaction survey and whatever the complaint, she would in turn send out an email dictating a new process that had to be implemented immediately! There was no due diligence to even substantiate the issue or to even consider the circumstances. It might have even been a one time occurrence. The executive's actions were reactionary at best and over reacting in most cases. IPI has an avenue for addressing issues without creating shallow unsubstantiated processes. IPI recognizes that there are sometimes best 'effective' practices that need to be implemented rather quickly.

For Executive Management, the magic is being able to discern the truth because there will be plenty of false messages mixed within the true process improvement message. There will be those that promise you that process improvement can fix everything that is wrong in the department. However, until IPI's workplace improvement, there is no evidence that traditional process improvement could begin to address the energy (QI) within the work environment. Being that the creators of IPI are among the people who do the real work, it was obvious there was an ingredient missing to create a truly wholistic solution (Intelligence, Surroundings (Ying Yang), Values (Feng Shui)).

IPI sits down with Executive Management to do some soul searching in order to find out the Executive Management style because this can dictate the implementation. Sometimes the root of the issue starts at the top. We find that some of Executive Management's images of their workers are often skewed. They don't really have a full picture and are ignorant to what the real environment and culture is within their company. Executive Management's impression of their workers is often based on the last mistake, the last disaster or failed project that has occurred. This is why process improvement implementations get off to the wrong start and get focused in the wrong areas. If the results from the gap analysis are prioritized by Executive Management who doesn't have a realistic picture of the day to day operations, then it's not the CMMI model that has failed nor can we entirely blame the consultant firms that try to implement the CIO's interpretation of the problem. Where they fail, IPI succeeds because we sit and hear the CIO's thoughts on how the department is running but then we work among the real workers for a period of time before we set out to fix anything. During our Environment Observation Phase (EOP), we aren't just looking at things with a critical eye, we are looking at it with an objective eye. We embed ourselves on the team to identify what is going well. Workplace improvement emphasizes 'foremost to do no harm' to the existing environment. That way, we are thinking about how to implement improvements that are not disruptive to the projects that are in progress.

We witnessed first hand how Executive Management was clueless about the culture within their software development department. They had a couple of workers that shared their opinion of the department and management took their word as gospel. Executive Management then hired a consultant firm and painted a picture of the department. The consulting firm was therefore defining changes based on third hand knowledge. Neither Executive Management nor the consultants could make good sound decisions because the basis of their knowledge was flawed. This is why the people doing the real work are the people running the IPI work place improvement implementation.

If management's mindset is not ready to embrace the full concept of workplace improvement, then it's sometimes best for IPI to turn down the job. If Executive Management is stuck on the ideology that they can best define the processes in order to push them down to the workers and if they cannot wholeheartedly embrace the IPI way of the 'bottom-up' approach, then surely we cannot expect the real workers to take to the vision.

Incremental IPI presents process in increments that can be mixed and matched. Other methods dictate one way to run your business because, face it - processes are a repeatable formula for running your business. You should be the one that comes up with a formula that makes your company unique. IPI doesn't feel like one-size-fits-all when it comes to processes.

IPI's Solution Set is a process improvement a la cart! IPI presents best practices from a variety of sources, you can pick some or you can skip some. You know what feels right for your organization. We also tell companies that over time what feels right might change.

We are not a cookie cutter solution. IPI solutions require creativity and some improvising. The final solution should be tailored made but at the same time contain ingredients that come from the fundamental practices used in the industry.

Like some other methods which we won't name, IPI doesn't send auditors to walk in your company and within 24 hours pass judgment on your organization because your company's processes don't fit their standardized method for processes. Auditors verify that you're doing the processes defined by the method collectively and consistently. Auditors are not judging whether those processes are providing a significant improvement to your organization. Organizations have to implement unnecessary processes because it's in the standardized method. IPI does not use a one-size-fits-all process improvement approach. IPI raises the question, "if every company small and large needs a process template that consists of 21 process areas?" Is that not a gross assumption that all companies are the same? How presumptuous is it to assume that all companies want to operate in the same manner? If that's the case, then how does a company establish dominance in the industry and produce superior products. To be a leader in your industry, you can't follow someone else's process template. Instead, the IPI process template is intended to be the starting point for tailored processes that create great ingenuity, not the definitive process standard. Auditing that needs to take place is internal. The number one question an IPI auditor would ask is whether the process met its quantitative goal?

In the IPI context, it is an incremental process buffet. IPI doesn't force feed you a one-size-fits all menu. IPI feeds in increments and gives you the right to refuse what doesn't work for you.

The buffet consists of our PCC 'People before Process' Infrastructure Outline, the FFMM Formula, The 3 Tier Bottom-Up Process Roll Out, and the SDLC Best Practice Buffet.

Of course, CMMI was well thought out and is tried and true. However, truth be told, some organizations somewhere is implementing processes because the CMMI book says so and not because there is a true need. That is why IPI encourages you to ask these questions:

◆ Should processes promote communication among resources?
◆ Should processes promote teamwork across the boundaries of departments?
◆ Should processes promote the vision of common goals?
◆ Are the metrics and results a way to confront reality through data processes to promote better decision making?

- Should processes reduce rework?
- Should processes promote money not being wasted?
- Should process metrics promote better schedules and cost estimation?

Measurably

"Metrics are a way to confront reality through data".

When you saw metrics, what you thought about was the typical measurements of a successful project – time, money, productivity, but IPI is anything but typical.

All other process improvement movements would simply measure data but IPI differentiates itself because we look at the data from all angles to achieve our wholistic approach (Intelligence, Surroundings and Values) to the work place. IPI feels that Harmony has to be measurably monitored. Processes should promote harmony and therefore it should be measured. Processes should reduce stress and that should be measured. Processes should promote Empowerment and therefore empowerment should be measured. If these factors are important to workplace harmony then they are important enough to baseline and track, don't just leave it up to chance. In EVERY process defined, the template guides you to define your Harmony, Reduced Frustration and increased Empowerment objectives and goals. To deliberately defining them, you will write your processes with the people who do the real work in mind. These important values and attributes of processes aren't just an after thought. For IPI, they are built into the templates. This is how IPI guides you to make your process improvement efforts accountable for incorporating Feng Shui. You are measurably accountable for these values. Therefore, Feng Shui is not just some novel reference or theme of IPI; it is a very real dimension.

How do you measure frustration, empowerment, harmony and other values of IPI and Feng Shui? First, define what these values mean to the people that do the real work. Perhaps it will have some similar characteristics to what we've defined in this section or maybe it means something different. Using the attributes of that definition weighs those values. Not only define them with words but ask them "what does frustration look like in your workplace?" "What does it sound like?" "What is frustrated behavior?" The answers may be something like: people call in sick more often, small things get escalated, lack of interaction (everyone just comes and sits at their desk), people don't want to participate in group functions and complaints are verbalized.

An analyst of values will utilize surveys, interviews, observations, shadowing and key lessons learned artifacts to extract and identify actions and attitudes that capture the environment and then baseline those attributes. Do a broad sampling as in surveying the targeted group daily for about 3 weeks. Note the highs and lows and find the median and the means. After the process is implemented, do your sampling over the course of the week. Not only see if there is positive change, but see if it's sustained and see that the high and low points start to level off. Do not be too severe but note if things stay at a constant level for long periods of time.

Another ideology that IPI has that is different is that we encourage you to make your metrics data transparent and available to the people doing the real work and to the customers. Something takes on even more value when it's brought out of the dark and into the light.

Target Improvement Phases (The left column being the initial target; the far right column being the ultimate target). Each organization should customize it and create realistic overall process improvement targets every 12 months.

Harmony	Extreme	Strong	Moderate	Medium	Mild
Frustration	Extreme A lot	Strong Often	Moderate Sporadically	Medium Infrequently	Mild Seldom
Empowerment	Extreme	Strong	Moderate	Medium	Mild
Rework	40%	20%	10%	6%	3%
Estimating	+/- 30%	+/- 20%	+/- 5%	+/- 3%	+/- 1%
Delivered Defects	X	½ X	¼ X	1/10X	1/100X
Defects Detected During QA Testing	30%	60%	80%	90%	99%
Productivity	X	1.5 X	2X	3-4X	>4X
Component Reuse	Negligible	Negligible	Occasional	30%	50%

If you get the part of IPI measurability, then the rest of the measurements fall in line with what those other process improvement initiatives have.

Measure the Project versus Measuring the Processes. Measuring the process in terms of the process being done and is it being done consistently.

Numbers can tell a story. Numbers take the mystery and guess work out of reporting progress. Every product and environment has unexpected surprises, but not every event in the project will be one of those unexpected surprises. Tracking the trends of your projects gives you the insight to start predicting some of the surprises. After you repeat the same thing on a regular basis, eventually you'll eliminate some of the surprises if the event is captured through processes and documentation.

Measuring projects and processes proves that you are implementing processes that are improving your environment that you have to measure before, during, and after the implementation. Metrics is a part of all process improvement methodologies. IPI's perspective incorporates measuring if the process is improving the harmony, the unity, as well as maintaining balance. IPI fundamentally wants to make sure that processes are not implemented to the extreme.

When the project is only focused on what the customer's wants, or meeting a deadline, or not exceeding the budget, or when the project's progress is being dictated by the developers, only then the project is out of balance. Project Managers often get blamed for the project, but a balanced project is determined from the onset based on the agreed upon processes. Without processes that create harmony, the Project Manager ends up spending their time tracking reactively rather than proactively as the project spins out of control. That is why IPI emphasizes just as every project has a communication plan, it should have a pro-active process plan and within that plan there should contain another plan for managing deviations from the agreed upon processes. The processes should define and contain everyone's interaction and accountability throughout the project lifecycle thus the processes making sure no one group that includes customers, Project Managers, Developers and Executive Management possess a one sided renegade influence over the project. There is a balance of power built into the project's process plan.

As everyone on the project races to the finish line, often the harmony, unity and balance gets trampled in the process. From the Feng Shui perspective - life is not about the destination, it's about the journey and so we apply this to projects. Everyone is focused on the deadline and the expected deliverables. Anything that gets in the way, like the users opinion or the reality of the current platform or the limitation of the employee resources is a 'problem'. Why not measure a project by how well it adapts to change, how quickly it absorbs the impact and reconfigures itself to refocus its efforts.

IPI encourages Harmony Dashboards that tracks, monitors and displays if all stakeholders are in harmony with one another. In other words, "are they on one accord?" When one is out of balance, there needs to be an audit and reconciliation taking place.

Added to the Project Manager's task is not only monitoring the budget or the timeline, but the harmony of the project.

Accountability

As the organization matures and more processes are established, it then has to police itself. This is where the accountability comes in. The PCC has to not only establish measurements for the project's work groups but it has to establish measurements for process improvement.

There are a whole set of measurements that have to be established and maintained so that there is a quantitative understanding of the performance of the organization's set of standard processes in support of quality and process-performance objectives, and to provide the process-performance data, baselines, and models to quantitatively manage the organization's projects.

Measurably Accountable
1. How do you measure that a process is being executed consistently?
2. How do you measure if a process is being followed consistently?
3. How do you measure if there is a predictable or sporadic trend occurring based on when the process is executed?
4. How do you measure that the process goal is being consistently met?
5. How do you measure if a process is performing and applied at its optimum?

Find, Fix, Measure and Maintain (FFMM – See Chapter 4 of Feng Shui for Intelligent Process Improvement ISBN# 9780-6151-4655-3) is part of the Process Hierarchy that provides tangible evidence that the process is working. The proof is in the numbers!

Harmony

When we first experienced a work environment of harmony and tried to define the key factors that contributed to the harmony, it was not easy to articulate. We started to recognize that process improvement was different from work place improvement. Process improvement focused on the projects and departments being successful and sometimes at the expense of moral. Whereas workplace improvement made us as workers feel good about our jobs and thus contributing our best. What we were feeling is what we identified as Harmony. After so many experiences in dysfunctional environments, it's now clear that the workers experience harmony when there is a clear understanding of their job. When they know what the expectations of the job are, they know what they are supposed to expect from others and they understand their relationship to their co-workers in the organization. When there are unknowns and ambiguity, there is disharmony. Processes define the 'rules of engagement' among the resources in the organization. Processes eliminate the unknowns. People aren't in fear of losing their jobs or being reprimanded or even fired. They are secure and at ease. Too many environments cause people to constantly look over their shoulders, feel insecure, feel inadequate, feeling defensive and non-productive.

We had a scenario where for 6 months, even though we were members of professional process improvement organizations, doing presentations, and writing magazine articles, we still became subject to the same feelings of inadequacy because the rules were constantly changing. You might get one assignment from one person via email and another assignment from another walking down the hall. If you wanted to take a day off, you had to copy no less than 5 people and put it on 2 different calendars.

We've been to several companies and the horror stories have the same ingredients: no organization, 11th hour assignments while juggling two other projects, no priority, upper management giving direct orders that contradicts your line managers expectations, your put on a project then taken off, hodge podge and unorthodox. You can't help but to be on edge. Emotions are high!

In no other software methodology would you find them putting an emphasis on Harmony. IPI identifies harmony as an essential goal for process improvement, thus reducing the frustration and headaches. Processes should not be disruptive to the flow of productivity. Processes should not make the people uneasy. If the process causes disharmony, it causes dysfunction and does not belong in the IPI organization.

Harmony is when order is restored physically and mentally. The process can dictate what you do physically but if your mentally not at ease with why your doing something, sometimes the thing that you are doing (i.e. the process) can be disruptive in the grand scheme of things. We studied and observed the phenomenon well enough until we realized that it was a very alarming pattern.

The real workers are IPI customers. The Process Hierarchy was included such that it incorporates all the people that directly and indirectly are impacted by the benefits of harmony in the workplace. The Process Hierarchy illustrates that if the workers that do the real work are harmonius, there is better work, better results, better project managers, better marketing, and more sells. The ultimate goal of upper management is to make the customer happy.

We introduced Feng Shui because it epitomizes the concept of living in harmony. We pointed out from the onset that Feng Shui is not introduced here as a religion. People have found all ways to tap into a source of what brings them inner peace and understanding. That power is what nurtures our soul or spiritual aspect. It's evident that whether you are practicing a faith or not, that we all have an inner, intangible aspect that also needs to be nurtured. That is why IPI identifies itself as a wholistic approach (i.e. Intelligence, Surroundings and Values) to workplace improvement – because we boldly proclaim that no other methodology actually focuses on the whole person – the mind, the body and the emotions of the people that do the real work! We address the mind by engaging their intelligence. We address the body when we refer to the Feng Shui practice. We appeal to the emotions as in those esoteric properties pride, ego, vanity, virtues, morals, principals and integrity. Do these things affect your work effort and your work outcome – unequivocally! We challenge anyone that can intelligently argue any different.

If you have pride in your work, then you'll produce better work. Have you seen where someone received negative criticism about their work to the point where they quit putting in the effort? Have you seen someone feel like their ego was attacked or that their integrity was being questioned? You can almost predict that that person is going to withdraw mentally and emotionally. Have you seen someone hold back doing something because vanity wouldn't let them take the risk of making a mistake? Have you seen where workers felt their virtues and principals were being breached for the sake of a project - management wanted you to lie, fudge numbers and hide the truth?

Processes promotes harmony by helping people feel good about themselves, their work and help people see the good in their fellow co-workers involved in the project.

Empowerment

Empowerment is about instilling accountability. Your first reaction might be that accountability is negative and has nothing to do with empowerment. IPI sees accountability as a positive. IPI implements accountability in such a way that it empowers people. From our experiences in software development, finger pointing is what people used as a defense mechanism when they have no power. QA blames Development for being late with their code. Development points the finger back at QA saying that QA's environment isn't ready for the code to be deployed and their test plans aren't created accurately. QA responds by saying the Business Analyst didn't give them the requirements on time and that the BA's also didn't give them good requirements. The BA's defend themselves by saying the end users didn't attend their JAD sessions. No professional likes to be accused of not doing their job is the moral of the story!

IPI shows how implementing accountability is a way for workers to acknowledge that they've done what is expected of them. It gives them the power to validate they've accomplished their job and made their contribution to the project. Process Improvement is a non-verbal form of tangible validation and verification of your contributions and achievements to the project.

IPI takes the guess work out of your progress. It gives you tangibles to support your case.

In a non-IPI environment, a typical scenario would be that we were on assignment and on a Monday upper Executive Management would come to us with a direct assignment to have an emergency release tested by Wednesday of that same week. From the moment we received the assignment we knew it was completely impossible even if we worked around the clock. In a world without metrics, all we had was our opinion against the CIO's directive, in our opinion "it couldn't be done". In the executive's opinion it could be done. With metrics, it's not about opinions, metrics provides the facts. Facts that both, the CIO and the workers has to acknowledge. Metrics is power. In this scenario, the metrics would have been proof that Wednesday was not possible. Metrics gives you the resources to support your case. Metrics and numbers are tangible powers that all members of the project including Executive Management can respect.

In scenario one, the real workers in IT have a voice and power as a result of tangible measurements.

A second scenario of empowerment based on accountability starts with another very familiar situation we encountered on our assignments where the QA is given a software release to test. The software release is in no way ready for QA. The most rudimentary unit testing has not been successfully tested. The QA department has to use their allotted schedule to do troubleshooting of the unit test failures. Therefore, the QA department is left with inadequate time to do real functional testing. This happened time and time again in various companies that were non-IPI environments. As a professional, it leaves you feeling defeated, like you have no say so. We felt like slaves. Whatever sloppy code they gave us, we had to test it to find out 90% of the code was terrible. We didn't have the option to reject, we just had to take patch work releases throughout the QA cycle. It was frowned upon to reject their release because we were causing the code to be rejected and the deadline being missed.

In one company in particular, we implemented IPI, the developers and QA came up with a process of accountability that related to the turnover of a release code from development to QA. The QA department had the power to reject the release code if 20% of the unit test did not pass. Quickly, Development started submitting better releases because they knew they would be held accountable. QA wasn't rejecting the release code arbitrarily and it was not personal, it was strickly process motivated. Without the process, QA didn't have the power and the developers took the rejection personal.

IPI's intent is for it not to be just about Harmony and Empowerment, but for Empowerment to equal Harmony. The more power that workers have over their environment and the more accepting they will be of it.

Many professionals initially resist accountability. They think it's like big brother watching over their shoulder. IPI circumvents the negative aspect of accountability. It's the workers that get to define how they are being held accountable. Having a voice in how you will be held accountable is what serving on the Process Champions Committee (PCC) is established for. Everyone and anyone can send a suggestion to the PCC, i.e. have an email box that accepts process improvement suggestions, each suggestion will be seriously considered and each suggestion will be deserving of a response.

Real Workers

The creators of IPI came through the ranks and wanted to stay true to their roots and to the people that we've been working with for 20 plus years! We wrote this book on behalf of all of the people that are the coffee-drinking, SUV driving, kid carpooling, American Idol watching, honor student bumper-sticking, mortgage-paying, drive through eating, t-shirt wearing, superstore shopping, electronic gadget overloading while doing the real work at companies around the world.

We are going out of our way to not be just about process methodology, we very much wanted to break the mold and if IPI had their way, we would be called 'a workplace improvement approach' for the real workers. Process improvement is just one component of IPI just as Feng Shui principals is a component as well.... but we decided that's a little too wordy! Another way IPI likes to state it is "Process Improvement is for the People and by the People!" We wanted to differentiate ourselves by putting the people before processes.

The people that do the real work are misunderstood. Management thinks they have to be in control on their own accord and that the workers lack the 'know how' and don't desire to create the best product even if they don't say Executive Managements actions make every worker feel as if they are lazy, shiftless, clueless, and incapable of making decisions. You can easily get an inferiority complex. Workers either get defensive or get defeated which is the heart of why they work in a constant state of frustration. Thank goodness for masseuses! If the IT industry never changed, then they wouldn't have any businesses. We worked in companies that had chair massages given twice a week at the company's expense – but did Human Resources ever consider doing some investigating to why the employees were so stressed to begin with.

When IPI refers to intelligent people, we equate thinking people with the working people. It's not to degrade anyone or to even say the executives are not working hard at what they do. But if anything, people create this perception that executives are the only ones capable of doing the thinking in corporate America and we want to dispel that stigma. The people that are working with their hands are equally capable of working and thinking at the same capacity if warranted.

Whatever happened to the cliché, 'the best teacher is experience?' That's why we are so adamant about stating that IPI is process improvement for the people and by the people.

When talking about the people doing the real work, we are not alienating anyone based on the IT role, responsibilities, salary or education. The people we are referring to are the people who have to get up at 6 am, get the kids ready for school, drive their 5 year old car through insane rush hour traffic. They then work their 8 hour days, eating leftovers from home out of a Tupperware dish, they go home to run the kids to sport practices, help the kids with homework, make a quick instant meal, wash the dishes, throw in a load of laundry, ignore the dog only to start getting ready for the next day of this endless routine. That is the image of someone married, it's the same image for the single people in IT who does the real work, except their routine doesn't involve the helping with homework and juggling sporting practice schedules, but instead there is usually a workout at the gym, juggling dates, and perhaps some television. The bottom line is that after a long day of work, the rest of the day is just fillers until it's time to go to work again. It doesn't take much to realize that such a significant part of working people's lives is spent at work. Do you want to spend that time being powerless and constantly stressed?

And to take it a step further, upper management has the misconception that the people doing the work are the cause of the flaws in the process resulting in bad software solutions. Don't be surprised if upper management's high level thinking concludes that the flaws in the process are because workers are lazy, which couldn't be further from the truth. Are the workers really working overtime, constantly stressed and juggling multiple priorities just to be cast as lazy?

The first premise of the IPI way of thinking is that the people want to do things in a better way, they want to do things the right way, they want to fix the flaws in the system and they have the knowledge to identify viable solutions. Our whole concept revolves around processes that serve a purpose that is of benefit to the people-hence the company. Process Improvement has had a history of self serving processes for the sake of processes that are simply overhead and not of any proven benefit.

The irony is IPI has observed that the thing that frustrates upper management is also the very same thing that frustrates the people doing the real work. Both entities want to produce something that they are proud of at the end of the day.

If you talk to the people that do the real work they will say that in fact they feel that "management won't let them produce their best". What managers don't realize by micro-managing workers is that they tie their hands and don't let them make pro-active decisions. By not letting people use their intellect, in turn makes their workers into the non-thinkers / robots they accuse them of being. If the manager hires a non-thinker, then they should recognize it early and terminate that individual's employment, but catering processes around one non-thinker, will demoralize the real thinkers among your workers.

Built into the IPI approach is giving the workers the ability to think about what they're doing as they're checking off their process checklist and doing process documentation for their projects. If and when a process doesn't serve its purpose, then IPI provides a built in whistle blower mechanism – to allow the organization to re-evaluate the process. This is the difference between a rigid, dictatorial process – driven method and a people-who-do-the-real-work driven approach. IPI wants processes that are natural and fit into the flow of the day to day work. When possible, create processes that document and account for themselves. This is where the balance of accountability fits in.

IPI implementation enables the least powerful workers to have the say so over the processes that determine their work day. IPI gives the workers ammunition not to use against the executives and decisions-makers but to use against the processes. IPI believes that processes that serve a purpose and improve the work environment should be adhered to, but processes that are broken that don't serve a purpose and are just overhead should be challenged and changed swiftly. IPI creates processes and an environment that levels the playing field between processes and the real workers. The processes don't control the people, the people control the processes.

We want processes that make people feel good about their job, their project, their product and their company. We want processes that don't create winners and losers. We want processes that make everyone a winner. Processes should never be a weapon to prove someone did something wrong or did not do their job.

Processes are to facilitate unity versus document the division among workers. Processes are to prove you did something right not that you did something wrong.

Like the example where development turned over a release that was not ready for QA, IPI doesn't create a process so that QA can "reject a release", the goal of the process is to communicate expectations between Development and QA of what is a baseline standard for the release to be turned over to QA.

The process should bring QA and Development together. They should define the process together with the same goal in mind, that goal toward harmony is to "establish expectations that each team can live with". In the end, the process should bridge the gap in the current way of turning over releases to QA. Through the usage of a common language of process improvement and process mapping both QA and Development can communicate. Processes also are a way to communicate and enlighten management of the work that is really involved in accomplishing the task. The new turnover process defined by QA and Development will expose a new level of integrity that needs to be introduced into the work flow that means an adjustment to timelines. Management will have to embrace the 'cliché' – "it takes time to do things right the first time". Visually understand the local level. Processes give management a view of the details which they typically have to remove themselves from.

We've talked about how IPI provides a common understanding of expectations between real workers in the same department, expectations for workers in different departments and expectations between workers and executives. That same understanding then filters to the customer. The customer should be aware of the processes necessary to deliver a quality solution and product. Customers are often frustrated with the timelines and a project plan is just a bunch of words. The process flow diagram is a visual representation that bridges communication between IT and the end users. The key we found from our experience is giving the customer several touch points along the way in the process. Let them 'peak under the hood' during the building process. Allow the customer to audit the processes, make your process documentation visible to them. It's like when you had to take a math test in school and the math teacher required that you 'show your work as you stepped through and worked through the problem to come to a conclusion'.

IPI creates a Software Development Ecosystem – in which all of the process roles co-exist in harmony, each carrying out their unique job function which contributes to quality software and satisfied customers. The 4 part Solution Set, Software Development Ecosystem and Process Hierarchy are the road map that defines order and purpose.

In the end, not everyone has to love or like each other, but they will have a clear understanding of their respective roles and responsibilities of what is expected. Isn't that a work environment you could appreciate and be appreciated?

After all, it's the real people that want to identify their flaws and be the ones to define the fix. It's like when someone talks bad about a family member and regardless of what they are saying be it bad or true, you feel you are the only one that has the right to talk about that family member. The people doing the real work feel like they are the only ones that have the right to criticize their work because their criticism takes into account those heroic efforts necessary to make the project happen.

Ecosystem (ē′kō-sĭs′təm)

[Scientific Ecosystem Definition]
A community of organisms together with their physical environment, viewed as a system of interacting and interdependent relationships and including such processes as the flow of energy through trophic levels and the cycling of chemical elements and compounds through living and nonliving components of the system.

[Software Development Ecosystem Definition]
A community of professionals together with their technical environment, viewed as a system of interacting and interdependent relationships and including such processes as the flow of positive and negative energy through the development life cycle.

Last but not least we want to elaborate on the word
Frustration.

Let's face it, IT is frustrating. In IT there are a lot of unknowns. As a matter of fact, there are no constants except the daily aggravations that the corporate culture of IT accepts as normal. Nothing is definite and environment remains dynamic. It's even more frustrating when there are no processes, but the worst case scenario is if the processes don't match the reality of the everchanging world of IT: deadlines change, assignments change , requirements change, customers change their minds, business needs change, technology changes, funding changes, resources come and go… so inevitability there needs to be a process implementation that lets you adjust your processes based on these changes.

To eliminate a lot of the frustration in IT, the processes have to reflect the many changes that go on in the real world. IPI was created with the full understanding that change is the nature of IT. IPI's incremental and customizable processes were created to accommodate change. We embrace changes. We predict changes. We are willing to adapt. Without change, our jobs wouldn't be exciting.

Merely creating processes doesn't restore the passion and the feeling of purpose. Harmony dictates more than just fixing what's wrong, it's about making the environment the best it can be.

We wrote this book because we were passionate in our thinking that the work place could be more than just bearable. Before we put a price tag on this book or even considered profits, we were committed to putting the vision of IPI on paper. We think that passion is what brings about radical and sustained change. Ultimately, once you implement IPI what you will find is a full transformation not just subtle improvement. We could have just accepted the status quo and left our frustrating jobs and desensitized ourselves by flipping through the TV channels and draining out the work day with can-sitcom-laughter and unreality-television, but it was indeed our passion that drove us to Borders every night after work until they announced over the loud speaker "We are now closed for the evening". We want people to implement IPI with the same passion in which we put into the defining and documenting of the concept.

The purpose of introducing Feng Shui in the work environment is to emphasize the importance of an emotionally positive work environment. A work environment that reconnects to the whole person. Its first task is refocusing executive leadership's attention to the morale issues and to find processes that address the work environment on a wholistic basis. Processes should nurture people's personalities, not stifle, not control, nor manipulate. Processes should be written to motivate the personalities within your work environment. It's about managing and creating processes from the heart with the people who do the real work in mind.

IPI How to…

Now for the practical implementation of IPI. Here is how you ensure you're implementing your process improvements the IPI way. That is what our IPI Implementation Checklist is here to show you:

How Does IPI relate to CMMI?
IPI is versatile and flexible enough to work with asny methodology of your choosing. IPI is an implementation approach. Once your PCC is setup, you can then implement CMMI

How Does IPI change the role and relationship between executives and the people that do the real work?
IPI believes that the people who do the real work are in the eye of the software development storm every day, so they experience the issues firsthand. Executives are only informed of those issues that are escalated to them. The IPI approach is such that the people who do the real work find the problem, fix the problem, measure it and maintain it. In this scenario, the Executive Committee is being informed of the solution in the form of a process rather than be informed about an escalated issue.

How do Business Analyst skills fit into the PCC?
Business Analysts are professionals at eliciting, gathering, managing, monitoring, prioritizing, analyzing, mapping and document stuff. That stuff is usually requirements but IPI has found that their skills work just as proficiently when it comes to processes.

How does the IPI Process Template help in keeping custom processes brief?
The IPI process approach is only guidelines but the message is to keep it simple, streamlined, and breezy. Our process templates direct you to keep your documentation short and straightforward. Here are some examples:
- When to use - process definition worksheets 1 – 2 pages
- When to create a glossary definition no more than 8 sentences
- When to create a checklist no more than 2 pages
- Department processes no more than 100 pages which includes a checklist
- Business local procedures 30 pages or less
- Project practices 2 –3 pages

How is it that IPI is very flexible however IPI feels that process mapping is a required artifact?
Process mapping is a visual roadmap of your processes. It can highlight gaps, identify points of contact between the workgroups and it shows you where you're going and where you are from beginning to end. We can't emphasize enough the importance of mapping your processes.

How does IPI guide you to not over document every best practice?
IPI's process portal definition template has a designated area for best practices, lessons learned and suggestions that are quite full fledged processes. This area can be called a best practice, effective practices, a process sandbox, process think tank or proposed process incubator. The information is stored and can be referenced and people may want to adopt them as local practices within their project. IPI has a saying – "there is a time and place for every process" but in some cases they don't need to be

How does IPI differentiate what level of management has to adhere to the processes?
IPI is emphatic that Executives and Management adhere to the process like everyone else. If they want to go around or ignore a process, they have to submit an Executive Process Waiver (EPW). The EPW gets documented and saved in the repository at the end of the project and wouldn't it be fun to see how many waivers were submitted by management and executives?

How can IPI claim not to be a Process Improvement One-Size-Fits-All solution, what makes them customized?
IPI provides templates and guidelines for process improvement, they set up a PCC made up of the people who do the real work, we do a comprehensive training on the four components Solution Set, we provide a generic template library, and from there each IPI implementation is different.

How does IPI see that everyone is involved?
On previous implementations, IPI has implemented an Online Suggestions Box on the Process Important Portal (PIP)

How does IPI incorporate process models?
Our standard new process definition template requires you to identify and update the process model.

How does IPI differentiate itself from Process Methodologies and Models?
Because IPI does not define what you should implement, it does define how you should implement the processes.

How does IPI ensure that the processes are not a one time event?
- ✔ The PCC is a permanent fixture in the organization
- ✔ Annual IPI boot camps help re-energize everyone's commitment to IPI

How does IPI keep from over processing (too detailed or too many processes)?
IPI believes that not every problem warrants a new process. IPI encourages the PCC to create a best practices sandbox on their process communication portal (PCP), where you can document some practices that work well or others may want to consider and items that don't warrant a process that becomes required. Secondly, IPI encourages the PCC to use the 3 Scenario Tests (3ST) on new processes. The Process has to be able to support three different project scenarios. That is one way to ensure it's broad enough, flexible and is not project specific. The piloting of new processes will also expose when a process is not worthy of being institutionalize or formalized. IPI encourages you not to 'rush to implementation' when institutionalizing new processes; take them for a long test drive through the PCC process piloting process.

How does IPI keep processes fresh and keep them from getting stale?
IPI encourages the PCC to put an expiration date on every process. The expiration date is part of the standard PCC process template. The expiration date reminds the PCC to review the process to make sure it is still meeting its goals and objectives based on the environment in case it has changed since the process was implemented. The librarian ideally should track the expiration date and notify the PCC when a process expiration date occurs.

How does IPI ensure that projects don't go and create project specific processes that are contrary to the policy, protocols or procedures?
IPI encourages the PCC to require that each project manager submits their Project Process Plan (PPP) for approval by the PCC steering committee.

How does IPI emphasize adapting to the change?
- ✔ The IPI process definition template is set up so that each step in a process can be identified as options, required or project defined. When the process is defined, the creators have to determine what is really required to meet the goal and then the other steps are highly recommended but leave it to the discretion of either

the project manager (who can make optional required by defining them as such in the PPP) or the actual person doing the real work.

✓ Submit an Executive Process Waiver (EPW) required whenever deviating from a policy, protocol, and procedural. An EPW form must be approved by executive management

How does IPI keep from being a one-size-fits all?
✓ Ultimately the PCC which is made up of the people in the company that do the real work, have total control over how they implement processes, PCC has no restrictions or limitations. Everything in this book is an IPI recommendation but not required.

How does an organization know if their doing IPI if everything is only a recommendation?
Three things will determine if you are doing IPI. 1) Implementing the IPI software ecosystem (including the PCC), 2) Process Hierarchy 3) The process hierarchy

How does IPI ensure every organization is improving processes in 90 days or less?
✓ IPI calls the process definition committee the champion's committee because it is made up of those people who understand, appreciate and promote process improvement. It is for the people who get it and get it first – the early adaptors. Then we leverage their enthusiasm to convert the naysayer. By the end of 90 days, you will have changed people's attitudes and only a few will remain resistant and they will be in the minority and for the most part be insignificant to the execution and implementation of IPI efforts.

How does IPI make process improvement so appealing?
IPI liberates the people who do the real work. They get liberated from always having to CYA or avoid the finger pointing that occurs when something goes from.

Why does the PCC only take 30 days but the first meeting is 90 days from the executive kick-off meeting in the IPI case study?
The Environment Observation Phase (EOP) takes 30 days. This is a realistic opportunity to observe and shadow the people that do the real work and to understand their corporate culture, the rules of engagement, and their informal processes. This is all done so that our recommendations are tailored made for each client. The case study is not a formula, it's a snapshot. Some IPI implementations take a lot less than 90 days but none have taken more than 90 days.

How does IPI address the 3 components of the wholistic concept?
We address and leverage the intelligence of the people by engaging their experience and knowledge to use it to be self governing and self defining. We address the surroundings and the environment by incorporating QI (negative and positive) energy. We address the values by assigning Feng Shui principles to the standing members of the steering committee.

How does IPI take process improvement to the next level?
Being that one half of the Lee and Kay duo spent significant years as a business analyst, the IPI approach was built on the foundation of a tried and true analytical approach to defining problems and defining solutions. In the case of IPI, the problem is with software development processes and the solution is improving those software development processes. The IPI templates encompass best practices from the business analyst profession, specifically using business scenarios (i.e. process scenarios), use cases, process flow diagrams, flow diagrams and event modeling. Time and time, we have gone to new client sites and their process documentation is a word document and they look like they tried to use every word in the dictionary to create verbose and voluminous documents which are not user friendly. IPI gets their clients to see that process analysis should be done by someone in the profession of eliciting and documenting such things.

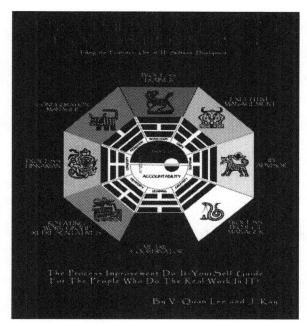

Feng Shui for Intelligent Process Improvement

By V. Quan Lee / J. Kay

This book is about giving the people who do the 'real' work the wherewithal and roadmap to making their workplace a creative, fun and pleasant environment and actually a place where a self aware person can thrive and be fulfilled in a balanced, wholistic way by incorporating Intelligence and Ying Yang values to a Feng Shui environment.

Title: Feng Shui for Intelligent Process Improvement
Author: V Quan Lee and J. Kay
Price: $99.95
Publisher and imprint: IIP LLP
Format: Paperback with full color interior pages
Number of pages in the finished book: 330
13-digit ISBN: 978-0-6151-4655-3
Month and day of publication: September 2007
Distribution arrangements: Bookmasters.Inc
Publicity contact information: IIP LLP a subsidiary of Ying Yang Enterprises

Feng Shui for Intelligent Process Improvements has received favorable reviews from professional organizations and academic leaders in the software development industry. The authors have been touring, consulting and lecturing on the topic for 3 years. They will be doing lectures through various organizations and colleges in 5 cities (New York, Atlanta, D.C/Maryland., Chicago, Seattle) before the release with the intention of adding additional appearances in other cities after the book is released.

About the Authors

V. Quan Lee is a ten year veteran, formerly trained in CMMI, an internal auditor at a level 5 organization and a senior QA resource that today applies his experience at establishing Process Champions Committees (PCC) utilizing Intelligent Process Improvement (IPI) methods at small and large companies that want to implement processes without being so rigid that they can't serve their customers needs. Mr. Lee is an alumnus of Morgan State University in Baltimore, Maryland where in 1996 he had the honor & privilege of being one of the few student employees to assist in implementing the overall infrastructure of the newly constructed Earl G. Graves School of Business & Management Computer facility prior to opening to the MSU public. Since graduating from MSU, he has worked in all facets of Information Technology in NJ, MD, DC, VA, GA, and most recently in Huntsville, Alabama ranging from the telecommunications to the state/local/federal government sector specializing in the Quality Assurance of Software and Processes. After 8 years of working and developing his skill-sets, he decided to chronicle the numerous software development processing faux pas with process improvements that he had encountered. As a result, he realized that there was a major disconnect between theory and application. Soon after, it became his personal goal to publish a book that clears up some of the common misconceptions of IT software process improvement. He holds a bachelors degree in Information of Science and Systems. He is an active member of SPIN (Software Process Improvement Network). Contact the author at vqlee@intelligentpi.com

About the Authors Continued

J. Kay (Sanders) is a Senior Instructor for B2TTraining.com, a premiere organization providing business analysis certification training for the International Institute of Business Analysis. She has been writing technical publications, coaching and training individuals throughout the country during her career in the areas of business analyst, database administration, data analysis and process improvement. She started her IT career twenty years ago upon receiving her degree in Computer Science from Bowling Green State University of Ohio. Her varied experiences have taught her how to adapt to the ever changing software development industry. She has accumulated a plethora of stories that she uses in the classroom and seminars to help bring to life software development best practices and pitfalls to avoid.

Her repartees in the software development industry, includes hospital information systems, banking, e-commerce, lottery, airline and the government sector. She has worked on full lifecycle projects, using various technical methodologies and models including Agile, Scrum, RAD, waterfall, object-oriented, UML, work flow automation and CMMI. She's held roles as project manager, program manager, tester, database administrator, developer, process analyst and business analyst. As a published author, she is currently completing her second book in a series of books related to Intelligent Process Improvement (IPI).

Jacqueline is as enthusiastic about learning as she is about teaching. Her teaching approach is to relate each lesson to real world experiences and identify how her students can apply it and reap immediate benefits. She has an interactive teaching style to insure the students leave the classroom energized and motivated to utilize the lessons learned. A member of IIBA, PMI and SPIN. Contact the author at jkay@intelligentpi.com

Tell Us What You Think.
We Would Love To Hear From You!
Let Us Know What Frustrates You About Software
Development/IT

Are you frustrated? Are you stressed out? Are you burnt out?
Are you ready to quit your job? Are you updating your resume
every few months? Are you from one job to another just to be
disappointed and disgruntled every time? Are you tired of all
those so called new methodologies that are going to make
everything all better like SDLC, CMMI, Six Sigma, IEEE, ISO, QMO,
PMI, Agile, Extreme, SPI? Tell us about it.. no really tell us... we
want to here your gripes. We also want to hear your ideas on how
an employee friendly IT shop should run so that it's not a sweat
shop of dysfunctional chaos. Tell us how you would run an IT
department or company if you had the power and were in the
position to do so. This group is here for you to vent your
frustrations to people who can identify with your pain so that we
will know that we are not alone. And that the way IT is run today
should be accepted as the norm.
IHateMyITJob@yahoogroups.com or
www.FrustratedNoMore.com

www.ingramcontent.com/pod-product-compliance
Lightning Source LLC
Chambersburg PA
CBHW051216050326
40689CB00008B/1331